Filipe Siqueira
Maria Elena Water

Corrosion inhibition in 1020 carbon steel plates

Filipe Siqueira
Maria Elena Water

Corrosion inhibition in 1020 carbon steel plates

Through the application of castor seed oil, aloe mucilage and ora-pro-nobis

ScienciaScripts

Imprint
Any brand names and product names mentioned in this book are subject to trademark, brand or patent protection and are trademarks or registered trademarks of their respective holders. The use of brand names, product names, common names, trade names, product descriptions etc. even without a particular marking in this work is in no way to be construed to mean that such names may be regarded as unrestricted in respect of trademark and brand protection legislation and could thus be used by anyone.

Cover image: www.ingimage.com

This book is a translation from the original published under ISBN 978-613-9-66154-1.

Publisher:
Sciencia Scripts
is a trademark of
Dodo Books Indian Ocean Ltd. and OmniScriptum S.R.L publishing group

120 High Road, East Finchley, London, N2 9ED, United Kingdom
Str. Armeneasca 28/1, office 1, Chisinau MD-2012, Republic of Moldova, Europe
Printed at: see last page
ISBN: 978-620-8-18157-4

CONTENTS

CORROSION INHIBITION IN 1020 CARBON STEEL PLATES THROUGH THE APPLICATION OF CASTOR SEED OIL, ALOE MUCILAGE AND ORA-PRO-NOBIS.

SUMMARY

Corrosion is a type of degradation that occurs in metallic materials, causing a loss of mass and altering the material's characteristics. This type of degradation occurs in the most diverse environments of our society, from household items to equipment components in companies. The most common resource used to reduce this type of damage is the use of corrosion inhibitors. However, most of these use zinc in their composition, which is extremely toxic both to the environment and to humans, even when administered in small quantities. Therefore, the study of natural corrosion inhibitors has been increasing. The substances responsible for inhibition present in plants are mainly p roteins and tannins. Castor bean oil and the mucilage of aloe and ora-pro-nobis are extracts that contain these substances in large quantities and can therefore be effective in combating corrosion. The aim of this study was to quantify the efficiency of using the three plants (castor bean, aloe and ora- pro-nobis) to inhibit corrosion in a saline environment. A resin was produced for this purpose, in addition to testing the extracts in natura, both in a saline environment. The best efficiency was found for ora-pro- nobis mucilage, due to its high protein content.

KEYWORDS: Corrosion, natural inhibitor, ora-pro-nobis.

1 INTRODUCTION

Corrosion is described as the degradation of a metallic material by physical, chemical or electrochemical processes in the environment. These procedures may or may not be combined with mechanical stress. The corrosion process takes place when oxidation reactions occur in metals, causing the loss of metallic material and thus altering their specific characteristics (FRAUCHES-SANTOS *et al.,* 2013). During the oxidation reaction, the metal acts as a reducer, donating electrons which are received by the oxidizing substance in the corrosive environment (GENTIL, 2007).

The most effective method of combating corrosion is to use inhibitors on parts subject to this condition. These inhibitors reduce or even prevent, in some cases, the reaction of the medium with the metal part, thus reducing the wear caused (FELIPE; MACIEL; MEDEIROS, 2013).During the manufacture of most inhibitors, the chemical element zinc is used in their formulation, which, due to the quantities used in the production of these corrosion inhibitors, is harmful to health and also to the environment (PERES, 2010).

Due to these factors, the demand for and research into natural corrosion inhibitors is growing. They have emerged as excellent alternatives, as they are produced from renewable sources, are biodegradable and relatively inexpensive. Most importantly, they do not contain heavy metals in their formulation (ALENCAR *et al.,* 2013). The function of organic inhibitors is to preserve the surface of the metal in the corrosive environment. To this end, a protective film is established in the area of metal-media interaction. Their effectiveness is subject to the quality of the film formed and it is therefore essential to know the physical and chemical characteristics of the plant used (PERES, 2010).

Research indicates that atoms of elements such as nitrogen, oxygen and sulphur, present in substances previously extracted from leaves, stems, roots, fruits and seeds, are effective in inhibiting the phenomenon of corrosion. However, inhibition occurs due to the presence of certain organic complexes present in the plant extract that are susceptible to oxidation. These include tannins, alkaloids, nitrogenous bases, carbohydrates, proteins and their acid hydrolysis products (ALENCAR *et al.*, 2013).

An example of a natural corrosion inhibitor is the leaf of the *Ricinus communis* plant, popularly known as the castor bean. Castor bean leaves are large, star-shaped with an average of eight points, can reach 60 cm in length and are toothed (AGEITEC, 2010).

Figure 1 - castor bean leaf.

Source: AGEITEC, 2010.

Castor bean leaf extract has already been used in tests to inhibit corrosion of steel in aqueous NaCl solution, as it is a medium that facilitates corrosion due to the increase in the conductivity of the medium, thus facilitating ion transport. The test results showed an 84% inhibition of corrosion in this material, as shown in Table 1 (ALENCAR *et al.*, 2013).

Table 1 - Corrosion efficiency of *Ricinus communis*

Data	*Ricinus communis* ("castor bean")
Concentration	3
Medium	00
Corrosive	PPm
Temperature	NaCl
Efficiency of	84%

Source: ALENCAR *et al.*, 2013.

Although castor bean seeds have a greater quantity of extractable oil, with characteristics very close to those of castor bean leaves, they have not yet been used to inhibit corrosion in metallic materials. Castor seed oil is known as ricin oil and is most commonly used as a lubricant, in cosmetics and also has some medicinal properties (CANGEMI, SANTOS, NETO, 2010). Castor bean oil can be obtained by chemical solvent extraction. To do this, the seeds can be crushed, which facilitates the extraction process as it causes the tissue and cell walls to break down, resulting in a larger contact area (CARVALHO, 2011).

Another plant not yet used for this purpose is *Aloe vera* (FIG. 2), also known as *aloe vera*. It is a plant from the liliaceae family of African origin, which spread to the American continent in the middle of the 15th century (STEVES, 1999). Aloe vera's flowers are shaped like a perianth made up of six parts, with no differentiation between the corolla and the calyx, and it has six stamens inside (STEVES, 1999). The outer part of its leaf is a bark approximately 2 mm thick, with a light green color, which, upon contact with air and light, begins to secrete a yellow resin (MARTINS, 2010).

It thrives in semi-desert or desert regions, but it also adapts to different regions (STEVES,

1999).

Figura 2   - Babosa. Source: CPT, 2015.

Around 1930, researchers began to identify the main compositions of aloe vera, which were aloin, emodin, chrysopanic acid, resin, gum, traces of volatile and non-volatile oils. Other studies found that the mucilage inside the leaf was mostly composed of various polysaccharides and also contained minerals, amino acids, vitamins, enzymes, tannins and steroids (STEVES, 1999).

Aloe Vera has analgesic, phytotherapeutic, detoxifying, energizing and nourishing properties and, in addition, the amino acids and tannins present in aloe are elements that can be effective in combating the inhibition of corrosion in metal parts (CURA PELA NATUREZA, 2013).The extract of the Aloe Vera plant is easily removed, by opening its leaves, you can already see a large amount of mucilaginous gel. You only need to press it to remove this liquid (MARTINS, 2010).

The ora-pro-nobis (FIG. 3) is also a plant with probable corrosion-inhibiting potential. The species' scientific name is *Pereskia aculeata* Mill. and it belongs to the cactaceae family. It is a type of rustic plant and is native to tropical America. It occurs in mesic or slightly arid regions (SOUZA *et al.,* 2009).

Figura 3 - Ora-pro-nobis.

Source: nossofuturoroubado, 2011.

This plant has generated a lot of interest, especially in the pharmacological field, due to its high protein and mucilage content. Studies show that 3.09g of protein and 0.75g of fiber can be found in every 100g of fresh material. It is considered to be the vegetable with the highest protein content (SOUZA *et al.,* 2009).

The anticorrosive action can be evaluated in two ways. The first is through the production of resin using the oil of the plant in its composition, according to a study by Vieira (2010).The second form of evaluation is also done for the oil and mucilage. The extract is used *in natura,* in saline solution, for later evaluation of their respective efficiencies, as described by Sathiyanathan, *et al.* (2005).

2 OBJECTIVE

The aim of this study is to compare the use of aloe and ora-pro- nobis mucilage and castor

seed oil as an anticorrosive for metal parts, assessing their respective efficacies.

3 METHODOLOGY

3.1 PRODUCTION OF CASTOR OIL-BASED RESIN

3.1.1 TEST SPECIMENS

To carry out the corrosion tests, two 1020 carbon steel plates measuring 10x5 cm and 3 mm thick were used as specimens. The specimens were previously sandblasted with silica microspheres to remove the oxide layer already present on them, so that the application of the extracts would not be compromised.

3.1.2 EXTRACTING OIL FROM CASTOR BEAN SEEDS

To obtain castor oil, the seeds were harvested from a plot of land in the Buritis neighborhood in the city of Belo Horizonte. The seeds were crushed using a blender. To extract the oil present in the seed, six samples of crushed seed with an average mass of 4.899g (+-0.020) were weighed on an analytical balance (Gehaka model B G400). The samples were wrapped in filter paper to form cartridges. These were tied with string.

The six samples were inserted separately into the Soxhlet, using hexane (150 mL) as the solvent. The extractions were completed after four days, for six hours a day. At the end of the extractions, 10 mL of castor bean oil was obtained. In order to try to increase the efficiency of the castor oil, a resin was made from it, following the methodology described by Vieira, (2010) presented below.

The efficiency of the castor oil-based resin will be compared with that obtained by Vieira (2010) in his tests using cashew nut liquid. The NP EN ISS04628 standard was used for this comparison.

They are referred to as "Ri", an abbreviation for *Rusting*. As such, the Standard presents six degrees of Ri ranging from Ri0 to Ri5, the lowest and highest levels of corrosion respectively.

3.1.3 PREPARATION OF RESIN BASED ON CASTOR SEED OIL

For the preparation of the resin, ricin oil (9.51g), formaldehyde (5 mL), phenol (11.97g) and oxalic acid as a catalyst (0.20g) were used as raw materials. The materials used in the preparation were a thermometer, a 250 mL flat-bottomed flask, a 1L beaker, a reflux condenser to prevent losses and a heating plate with magnetic stirring. The temperature of the mixture was controlled by the temperature of the water in which the flask was immersed, as shown in Figure 4.

Figura 4   - Preparation of castor oil-based resin.

The phenol was first added to the castor seed oil. The mixture was stirred until it reached a temperature of 75°C at room pressure. Afterwards, the catalyst, oxalic acid, was added and waited until it was completely homogenized in the mixture. Formaldehyde was then added. The reaction time was two hours. After this time, the resin (FIG. 5) was cooled to room temperature and finally applied to the entire test specimen.

Figura 5 - Castor oil-based resin ready.

3.1.4 TESTING SPECIMENS WITH CASTOR OIL RESIN

After application, the test specimens were given a four-day period in which to completely cure the possibly anti-corrosive layers. They were all sent for testing in salt spray equipment, the purpose of which is to simulate what normally happens in areas with this type of atmosphere, i.e. the presence of salt and high humidity. The tests were carried out at the FIAT company, located on highway BR 381, km 429 in *Betim* - MG. The test took place over seven days for all the plates.

The test was carried out according to ASTM B117: 2009 adapted to ITA Standard 50180:2011 - Item 2.5.6.1, within the following parameters: temperature 35°C, NaCl concentration of 50 ± 5 g/L, lasting 168 hours, following the methodology described by Vieira (2010).

The equipment on which the tests were carried out (FIG. 6) is Bass Equipamentos LTDA, model CCT- GS-STD-i-04/2012.

Figure 6 - Salt spray equipment. Source: Own collection, 2015

3.2 CASTOR BEAN EXTRACT, ALOE VERA AND ORA-PRO-NOBIS IN NATURA

3.2.1 SPECIMEN PREPARATION

For this procedure, four 1020 carbon steel plates measuring 5X2 cm and 1 cm thick were used as specimens. The specimens were polished with 320 and 180 grit NORTON sandpaper in order to remove the already oxidized layers of the metal. Trichloroethylene was then used to degrease the plates.

3.2.2 PREPARATION OF CASTOR BEAN, ALOE AND ORA-PRO- NOBIS EXTRACTS

The castor bean seed oil was extracted as described in section 3.1.2. To remove the mucilage present in the pulp of the Aloe Vera plant, two leaves of the plant were cut open with a knife and the pulp was removed manually. This extract was placed in a 50 mL beaker.

The *Ora-pro-nobis* extract was obtained by macerating the leaves using a grater and pistil. As the maceration is carried out, the plant's mucilage is gradually released from the leaves. This mucilage was stored inside the grater itself.

The efficiency of aloe and *Ora-pro- nobis* mucilage and castor seed oil used as anticorrosives was compared with the efficiency of castor bean leaf used for the same purpose, as shown in a study by Sathiyanathan, *et al.* (2005). In this study, the specimens were subjected to a corrosion test in a saline environment. After the test, the efficiency of the corrosion inhibitor is calculated by comparing the loss of metal mass in the presence of the inhibitor and in its absence. Efficiency is therefore measured in terms of mass per unit area.

We also took into account what was described by Alencar, *et al.* (2013), for an extract to be efficient it needs to reach 70% or more in inhibiting corrosion. After all the extracts were removed from the plants, 300ppm of them were used to pass through the test specimens, which were prepared properly for the tests, and the test media could then be prepared.

3.2.3 PREPARATION OF TEST MEDIA

The study carried out by Sathiyanathan, *et al* (2005), in which the best corrosion inhibition was achieved at a concentration of 300 ppm of the extracts, was taken as a starting point. 200mL of 100 ppm sodium chloride solution was prepared and divided into four different beakers. The castor bean, aloe and ora-pro- nobis extract solutions were placed in the first three, and the last beaker was used as a blank.

The previously polished and degreased specimens were completely immersed in the beakers containing the 100 ppm NaCl solutions. The test took place over a period of four days.

3.2.4 MASS LOSS METHOD

Weight loss measurements were taken every day during the four days of testing, at the same time of day, so that in addition to studying mass loss, the corrosion rate of the parts could be studied. At the end of the four days of testing, the specimens were removed from the solution, washed with distilled water and then with trichloroethylene and the last weighing was carried out. The analytical balance used to weigh the specimens was a Gehaka model B G400.

The percentage of inhibition efficiency was calculated using Equation 1.

$$P = 100 * (1 - W2/W1) \qquad (1)$$

Where P is the percentage of inhibition efficiency; and W1 and W2 are the corrosion rates in the presence and absence of the so-called inhibiting concentration (SATHIYANATHAN *et al.*, 2005).

3.2.5 ELECTROCHEMICAL STUDY

The electrochemical potentials of the NaCl solutions, at 100ppm, with their different extracts were measured during all four test days, at the same time, using a digital multimeter. The variation curves of the potential versus exposure time for the data obtained have been plotted in Graph 1.

Metal parts normally dissociate into cations when in contact with an electric current. The cations dispersed in contact with the electrolyte solution, due to oxidation-reduction reactions, are deposited back onto the surface of the parts, thus causing them to corrode. The higher the energy, the larger the deposited layer becomes, i.e. the lower the current measured by the multimeter, the less corrosion there is in that medium (MARCOS, NUNES, 2003).

3.2.6 CORROSION RATE STUDY

The corrosion rate can be determined in several ways, the most common of which is the loss of mass per unit area per unit time. By measuring the mass of all the specimens before immersion in the sodium chloride solution and then removing them once a day at the same time to weigh them over the four test days, the corrosion rate can be assessed. The value of the total mass loss at the end of the time was divided by the time the tests lasted, thus giving us an idea of the average speed (rate) of corrosion.

4 RESULTS AND DISCUSSION

4.1 CASTOR OIL-BASED RESIN

The specimens were weighed before and after applying the resin and again after the test period. The mass values are shown in Table 2.

Plate	Masses (g)		
	Initial	Final	Loss
01	12,988	12,424	0,564
02	12,876	12,39	0,486

Table 2 - Specimen masses.

According to the data in Table 2, a considerable loss of mass can be seen for both resin-applied parts. Based on the NP EN ISO 4628 standard, the specimens with the resin applied were classified as Ri5, according to the visual assessment of the degree of rust/corrosion.

Also according to this standard, for a resin to be considered efficient, its rating must be below Ri4. When it is equal to or above this, it is necessary to repair or even replace the material. According to Vieira (2010), the resin produced with cashew nut shell liquid obtained a result of Ri3. According to the author, the good efficiency of this resin is due to a compound present in cashew nut shell liquid, cardanol (FIG. 7).

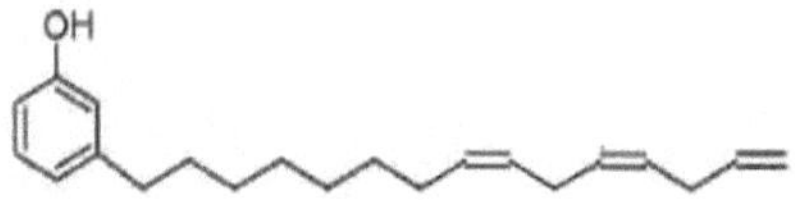

Figure 7 - Structural formula of cardanol.

Source: MAZZETTO; LOMONACO; MELE, 2009.

This compound is a source of phenol, which occurs naturally in this almond, and can therefore replace it in various applications. One of them is anti-corrosion resins, and in some cases the results are better.

The inefficiency of the resin based on castor bean oil is probably due to the fact that this seed does not contain the compound cardanol. When cardanol is present in the reaction medium, a smaller amount of phenol is used to produce the resin. Therefore, it is likely that the amount of phenol used in the manufacture of the resin was not sufficient for its efficiency.

4.2 CASTOR BEAN EXTRACT, ALOE VERA AND ORA-PRO-NOBIS *IN NATURA*

4.2.1 METHOD FOR MASS LOSS

The results obtained for the variation in the loss of mass of the specimens after four days of immersion in an aqueous solution of NaCl at 100ppm, as well as 300ppm of *in natura* extract of each sample, are shown in Table 3.

Table 3 - Loss of mass of the specimens.

	Mass (g)				
Extract	Day 01	Day 02	Day 3	Day 4	Loss total
Seed of Castor beans	8,316	8,314	8,284	8,258	0,058
Babosa	8,524	8,522	8,482	8,463	0,061
Ora-pro-Nobis	8,521	8,520	8,519	8,495	0,026
Solving White	8,730	8,680	8,601	8,576	0,154

According to the data presented, mass loss occurred for all the extracts, the difference being in the amount of loss. As expected, the greatest loss of mass occurred in the blank specimen, since it did not have any type of protection. The lowest mass loss was seen in the ora-pro-nobis extract, with 0.026g. The results obtained by Sathiyanathan, *et al.* (2005), show an even lower mass loss using castor bean leaf extract, 0.024g. Using the mass losses and Eq. 1, the percentage inhibition efficiency of each extract used can be calculated. The results of the efficiencies are shown in Table 4.

Table 4- Corrosion inhibition efficiency of castor seed oil and aloe and ora-pro-nobis mucilages

Extract	Corrosion rate (g/h)
Seed of	0,0006
castor bean Babosa	0,0006
Ora-pro-nobis	0,0003

The best corrosion inhibition efficiency was found for ora-pro- nobis mucilage (83.11%).

4.2.2 ELECTROCHEMICAL STUDY

The electrochemical behavior in each solution with the different extracts is shown in Graph 1, which was plotted using the daily measurements shown in Table 5.

Table 5 - Electrochemical potential of the samples.

Extra to	Electrochemical Potential (mV)			
	Day1	Day2	Day3	Day4
Castor bean seed	-373	- 20 6	- 13 2	-129
Babosa	-552		-32	13

Ora-pro-nobis	-262	70		-157
White solution	-212	19 7 19 5	17 1 24	155

Graph 1 - Electrochemical potential of the samples.

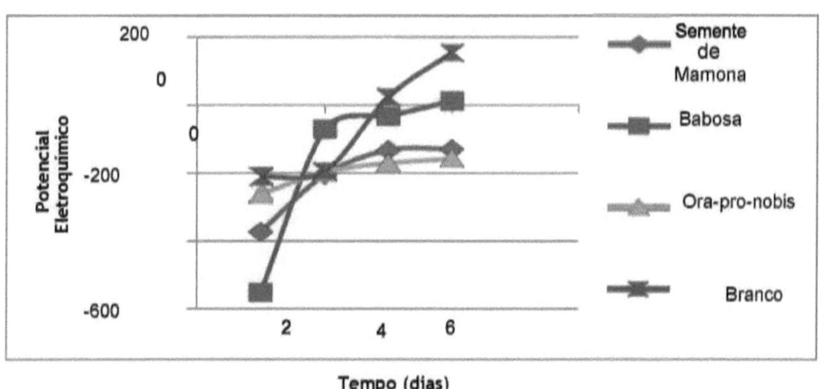

The extracts used may not be sufficient to form a stable corrosion environment. However, according to the tests, they all performed better than the one without any type of extract. The extract that obtained the best results for the electrochemical study, as can be seen in Graph 1, is the ora-pro-nobis extract, with values ranging from -262 mV to -157 mV. The studies carried out by Sathiyanathan, *et al.* (2005), show an 84% corrosion inhibition efficiency for the castor bean leaf extract, with electrochemical potentials ranging from -720 mV to -271 mV. Even though the difference in potential is very large between the study by Sathiyanathan, *et al.* (2005), and the ora-pro-nobis mucilage, it is still satisfactory due to its low loss of mass and an efficiency of over 70%.

4.2.3 CORROSION RATE STUDY

From the mass loss measurements, Graphs 2, 3 and 4 of mass loss as a function of time could

be constructed to give an idea of the corrosion rate for each of the extracts.

Graph 2 - Corrosion rate of 1020 carbon steel plate with castor seed oil.

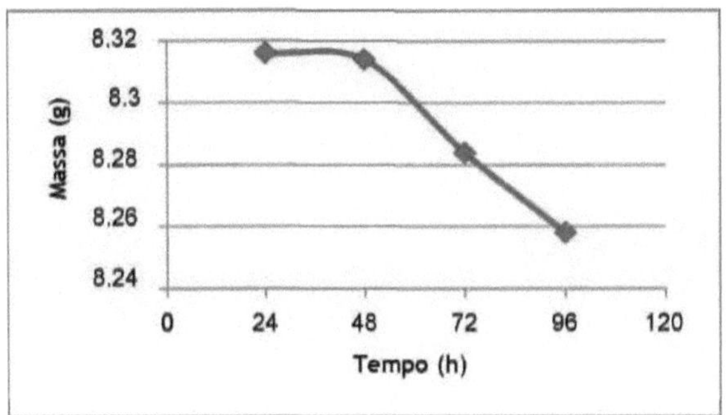

Graph 2 shows that the mass loss for castor bean extract increased with time. At the beginning

the mass of the plate was 8.316g at the end of the tests, i.e. after 96 hours the mass of the 1020

carbon steel plate was 8.258g.

Graph 3 - Corrosion rate of 1020 carbon steel plate with aloe mucilage.

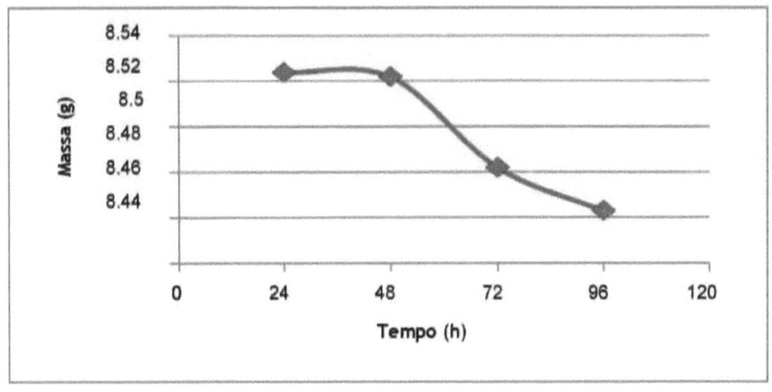

Graph 3 shows the loss of mass in NaCl solution with aloe extract. You can see a similarity between graphs 2 and 3, as the total mass loss over time was very close (0.058g for castor bean extract and 0.061g for aloe extract).

Graph 4 - Corrosion rate of 1020 carbon steel plate with ora-pro- nobis mucilage.

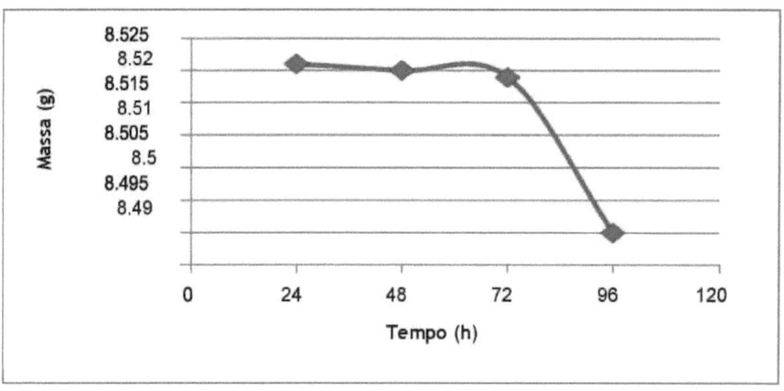

Graph 4 shows a smaller loss of mass over time than the other extracts. At the beginning, the specimen weighed 8.521g and at the end of the 96-hour test, the specimen weighed 8.495g, showing a total loss of mass of 0.026g.

After plotting the graphs, the corrosion rate can be calculated. The values found are shown in Table 6.

Table 6 - Corrosion rate of 1020 carbon steel plates with the extracts *in natura*.

Extract	Efficiency (%)
Castor bean seed	62,34
Babosa	60,41
Ora-pro-nobis	83,11

The corrosion rate is lower for the ora-pro-nobis extract (Table 6), while the castor seed oil and aloe extracts showed similar and higher values, showing that corrosion occurs more quickly in the presence of these two extracts. As well as the corrosion rate being lower in the presence of ora-pro-nobis mucilage, the mass loss tests, corrosion inhibition efficiency and the electrochemical study also showed that this plant has a better potential for inhibiting corrosive processes.

5 CONCLUSION

According to the NP EN ISO 4628 standard, the castor oil-based resin was not efficient, as it is classified as Ri5. The production of an anticorrosive resin using renewable sources is not efficient with any plant extract. An in-depth study should be made of the characteristics of the extract as well as the components needed for this resin to be efficient. For example, the use of plants that contain the compound cardanol, which can be used for

to replace part of the phenol in the manufacture of resins.

For the *in natura* extracts, the maximum inhibition efficiency obtained was with the *ora-pro-nobis* mucilage (83.11%), at a concentration of 300ppm of inhibitor in a 100ppm NaCl solution. The current values measured by the multimeter decreased with time for all the samples. However, the most negative value remained in the *ora-pro-nobis* mucilage sample, with -157mV. Therefore, corrosion occurred to a lesser extent for this sample. As well as being biodegradable and environmentally friendly, this extract has inhibitory properties for protecting 1020 carbon steel. This can be attributed to the large quantities of proteins present in the extract, as these are extremely important substances in inhibiting corrosion. Aloe mucilage and castor seed oil cannot be characterized as efficient, as their values for this ratio were less than 70%.

ANTICORROSIVE ACTION OF ORA-PRO-NOBIS MUCILAGE ON CARBON 1020 STEEL PLATES

SUMMARY

Corrosion is a type of degradation that occurs in metallic materials, causing a loss of mass and altering the material's characteristics. This type of degradation occurs in a variety of environments.

The most commonly used corrosion inhibitors are used in all areas of society, from household items to components and equipment in companies. The most commonly used resource to reduce this type of damage is the use of corrosion inhibitors. However, most of these use zinc in their composition, which is extremely toxic to both the environment and humans, even if mixed in small quantities. Therefore, the study of natural corrosion inhibitors has been increasing. The substances responsible for inhibition present in plants are mainly protein and tannins. The higher the crude protein content, the better the anti-corrosion action, so the protein test was carried out to check the content of each species studied. The aim of this study was to compare the use of resin, prepared with mucilage from two different species (Pereskia aculeata Mill and Pereskia aculeata var. godseffiana), with thorns and without thorns respectively, from ora-pro-nobis, and mucilage in natura, evaluating the protein content and anticorrosive properties of the same for 1020 carbon steel.This work showed that the fresh mucilage obtained satisfactory results, with the ora-pro-nobis without a thorn having an inhibition efficiency of 86.75% and 93.57% with a thorn. However, the use of an anti-corrosion resin produced from renewable sources is not efficient with the mucilage of these plants.

Keywords: corrosion, mucilage, ora-pro-nobis, 1020 carbon steel, natural inhibitors, anticorrosives.

24

1 INTRODUCTION

Carbon steel is the most widely used metal alloy for building structures and equipment due to its considerable mechanical properties. "Carbon steel" is generally defined as an iron alloy containing between 0.05 and 2.0% carbon by mass. In addition to iron and carbon, these steels always contain some amount of manganese, sulphur and phosphorus, and may also contain small amounts of silicon, aluminium and copper. As iron is considered an active metal, it is almost always necessary to apply a corrosion prevention method, usually in the form of a metallic or non-metallic, organic or inorganic coating (BOSSARDI, 2007).

The formation of rust is one of the most unpleasant processes that man witnesses on a daily basis, knowing that it will inevitably lead to the repair or replacement of metallic material. This phenomenon occurs when ferrous alloys such as carbon steels are exposed to the atmosphere or submerged in water. These tend to react with the corrosive environment to form stable compounds and, therefore, the metallic material is lost and its properties are altered. The corrosion process takes place when oxidation reactions occur in metals, where the surface becomes corroded (OLDHAM; MYLAND, 1994).

In many industries, corrosion effects can complicate the selection of suitable materials for certain applications. Currently, improving corrosion resistance is only achieved through an increase in cost. However, the actual cost of the materials used in a project depends on the corrosiveness of the environment (RAHIM, 2011). Equipment that is well protected against external aggressions avoids unnecessary downtime for maintenance, preventing accidents and damage to the equipment (PERES, 2010).

The application of anti-corrosion strategies can slow down or practically eliminate the oxidation reaction. The use of corrosion inhibitors is very popular, *as* they can be applied in various sectors of industry, such as cooling systems, oil refinery units, acid pickling baths and gas production units. These inhibitors are substances which, when added in small concentrations to the corrosive medium, have the purpose of preventing or reducing the reaction of the metal with the medium (GOMES *et al.*, 2005).

Most inhibitors work by adsorbing ions or molecules onto the surface of the metal. However, there are some considerations that must be taken into account when choosing an inhibitor, such as cost, which can be very high when the material involved is expensive or when the quantity required is large; toxicity of the inhibitor, which can cause dangerous effects on humans and other living species; availability of the inhibitor will determine its selection. Although many synthetic compounds exhibit good anti-corrosion activity, most are highly toxic to living beings and the environment. The safety and environmental problems of corrosion inhibitors used in industries have been questioned and received attention globally. Toxicity can manifest itself either during the synthesis of the compound or during its application (GOMES *et al.*, 2005).

Toxic effects have led to the use of natural products as anti-corrosion agents that are ecologically friendly and less harmful. Today, many alternative corrosion inhibitors have been developed, ranging from herbal plant extracts to essential oils from parts of the plant.

of plants originating from different geoclimatic regions of the planet (SHARMA; MODHOO, 2012).

According to Standard NP EN ISO 4628 (CIN. 2011), the assessment of the degree of rusting corresponds to the area with visible oxidation of the surfaces. The degrees of corrosion are referred to as "Ri". Table 1 shows the areas for each level of corrosion.

Table 1- Evaluation of the degree of oxidation - Rusting

THIS	Oxidized area (%)
Ri0	0
Ri1	0,05
Ri2	0,5
Ri3	1
Ri4	8
Ri5	40/50

Source: CIN, 2011

New studies are being carried out on the exploitation of natural products obtained from plants as corrosion inhibitors, which are ecologically acceptable, cheap, readily available and renewable sources of materials. Extracts from leaves, branches, stems, roots and fruits comprise a mixture of organic compounds containing nitrogen, sulphur and oxygen atoms, have been studied and act as effective corrosion inhibitors in different aggressive media (GUNAVATHY; MUROGAVEL, 2012).

The corrosion inhibition efficiency of these extracts is usually due to the presence in their composition of complex organic species such as tannins, alkaloids and nitrogenous bases, carbohydrates and proteins, as well as their acid hydrolysis products (SATAPATHY *et al.*, 2009).

Tannins (compounds of plant origin, consisting of simple carbohydrates, hydroxycolloidal gums, phenols and amino acids) have been studied over the last few decades. Their inhibitory action is due to their reaction with iron oxide, forming an insoluble iron-tannate complex. Among the various applications of tannins in the study of corrosion is the formulation of anticorrosive primers and corrosion inhibitors (PERES, 2010).

Tanning substances obtained from plants, whose characteristic properties consist of adsorbing metals dissolved in water, due to their color and viscosity, can then be successfully used in the anticorrosive, flocculant, beverage and plastics industries, etc (MARTINEZ, 1996).

The classes of natural compounds with high antioxidant potential are phenolic acids, flavonoids and polyphenolic compounds. The presence of phenolic groups in these compounds promotes the stabilization of free radicals due to the resonance structures that can be formed, which is why they have high antioxidant activity (GHASEMZADEH; GHASEMZADEH, 2011).

Among the species of the Cactaceae family, the Pereskia genus has shown that some individuals have compounds with antioxidant activity, such as the fruits of *Pereskia aculeata Mill* (ora-pro-nobis) (AGOSTINI-COSTA *et al.*, 2012). The ora-pro- nobis is a type of rustic plant and is native to tropical America. It is found in mesic or severely arid regions. This plant has generated a lot of interest, especially in the pharmacological field, due to its high protein and mucilage content. Studies show that 3.09g of protein and 0.75g of fiber can be found in every 100g of fresh material. It is considered to be the vegetable with the highest protein content (SOUZA *et al.,* 2009).

There are several species of ora-pro-nobis, among them: *Pereskia aculeata Mill* (called common ora-pro-nobis, white flowers), *Pereskia aculeata var. godseffiana* (golden ora-pro-nobis), *Pereskia bleo* (orange ora- pro- nobis) and *Pereskia grandifolia* (pink ora-pro-nobis). Two of these were studied in this work, *Pereskia aculeata Mill and Pereskia aculeata var. godseffiana.* According to Sathiyanathan *et al.* (2005), the corrosion inhibition efficiency of mucilage is mainly due to the presence of protein, i.e. the higher the crude protein content, the better the anti-corrosion action. In view of the need for anti-corrosion inhibitors from renewable, biodegradable sources, with relatively low economic importance and less unhealthy, with ora-pro-nobis being a probable potential corrosion inhibitor, a study of the action of this type of plant as an antioxidant on 1020 carbon plate is justified. In view of the above, the aim of this study is to evaluate the anti-corrosion properties of two species of ora-pro-nobis - the mucilage *in natura* and in resin - on 1020 carbon steel.

2 METHODOLOGY

All the tests were carried out in the Organic Chemistry, Instrumental Chemistry and Bromatology laboratories at the University Center of Belo Horizonte (Unibh). The methodology was based on the study by Sathiyanathan, *et al.*, (2005) and the protein analysis methods of the Adolfo Lutz Institute (2008).

2.1 MATERIALS AND REAGENTS

Equipment: Analytical balance, Digester block, Magnetic stirrer, Chapel. Reagents: Hydrochloric acid (Synth), Sulphuric acid P.A (Dinàmica), Oxalic acid (Dinàmica), Methylene blue (Vetec), Phenol (Synth), Formaldehyde (Dinàmica), Sodium hydroxide (Neon), Anhydrous copper sulphate (Synth); Anhydrous potassium sulphate (Nuclear) and Methyl red (Vetec).

2.2 PROTEIN TEST

The plants used for the tests were obtained from home cultivation in the south-central region of Belo Horizonte. Both species were used for the tests and identified as: *Pereskia aculeata Mill* with thorns and *Pereskia aculeata var. godseffiana* without thorns. According to IAL (2008), the principle of the classic Kjeldahl method consists of transforming nitrogen from nitrogenous substances by boiling with concentrated sulphuric acid and catalysts into ammonium sulphate. To determine the protein content of Ora-pro-nobis, 0.3 grams of the sample were weighed out on paper. They were then transferred to the digestion tubes, adding about 1 g of the catalyst mixture (copper sulphate and potassium sulphate) and 5 mL of sulphuric acid.

The samples were then heated in a digestion block in the chapel, starting the digestions at 50°C and increasing the temperature by 50°C every 1 hour until it reached 350°C. The digestion of the samples, carried out on both species, was completed when the inner walls of the tubes were clean, without white fumes, and the emerald green liquids were clear or colorless.

The digestion tubes containing the digested samples were immediately transferred to the distillation set. Using an addition funnel, 40 mL of 50% sodium hydroxide solution was added to the tube containing the digested sample until a slight excess of base was obtained, releasing the ammonia in the form of ammonium hydroxide, which was distilled into approximately 100 mL and collected in 20 mL of 2% boric acid solution and Anderssen's indicator (methyl red and methylene blue). Nitrogen was then determined by titration with 0.02N hydrochloric acid until the color of the solution (bluish-green) turned purple (light purple). The crude protein content (proteins per cent m/m) of Ora-pro- nobis was calculated using Equation 1.

(1)   Equation 1:

$$T = \frac{V \times N \times 0,14 \times f \times 100}{P}$$

Where:

V = volume of hydrochloric acid in mL;

N = normality of the acid solution

in mol / liter or N; P = mass

of the sample in grams;

f = conversion factor (6.25, IAL, 2008) T= crude protein content

2.3 MUCILAGE PREPARATION

The Ora-pro-nobis extract was obtained by macerating the leaves in a mortar and pestle (Figure 1). As the maceration takes place, the plant's mucilage is gradually released from the leaves.

Figure 1 - Maceration of Ora-pro-nobis to obtain mucilage.

2.4 PRODUCTION OF RESIN BASED ON ORA-PRO-NOBIS OIL

In order to increase the efficiency of ora-pro-nobis mucilage, a resin was produced using it. The resin was prepared using:

1.	Ora-Pro-Nobis with thorn: extracted mucilage (9.58g), formaldehyde (5 mL), phenol (23.95g) and oxalic acid as a catalyst (0.209g);

2.	Ora-Pro-Nobis without thorn: extracted mucilage (9.57g), formaldehyde (5 mL), phenol (23.93g) and oxalic acid as a catalyst (0.216g).

First, phenol was added to the Ora-pro-nobis mucilage. This mixture was heated under stirring conditions to approximately 75°C under ambient pressure. Then the catalyst (oxalic acid) was added and the mixture was allowed to homogenize completely. The formaldehyde was then added and the resin was allowed to cool to room temperature before being used.

2.5 APPLYING MUCILAGE TO PLANTS

After removing all the mucilage from the plant, a thin layer of it was applied to the test specimens, which were properly prepared for the tests.

2.6 TEST SPECIMENS

Sixteen 1020 carbon steel plates measuring approximately 10x5 cm and 3 mm thick were used as test specimens for the corrosion tests. The specimens were sandblasted to remove the already oxidized layers of metal. After applying the resin or mucilage (separately), the specimens were left for a period of four days for the possibly anti-corrosive layers to cure completely.

2.7 TEST MEDIA PREPARATION

According to Sathiyanathan, *et al.* (2005), the best corrosion inhibition is at a concentration of 300 ppm of the extracts. This was the concentration used. 1000mL of 100 ppm sodium chloride solution was prepared and divided into sixteen different beakers, identified with numbers from 1 to 16 (Table 2). The tests were carried out in triplicate.

Table 2 - Identification of the beakers used for mass loss

Beakers	Test Body
1	White
2,3,4	Thornless ora-pro-nobis mucilage
5,6,7	Thornless ora-pro-nobis mucilage with resin
8,9,10	Ora-pro-nobis mucilage with thorns
11,12,13	Ora-pro-nobis mucilage with thorn and resin
14,15,16	Resin (control)

The previously prepared specimens (1 to 16) were completely immersed in the beakers containing the 100 ppm NaCl solutions (Figure 2). The test took place over a period of four days, with daily evaluations of mass loss.

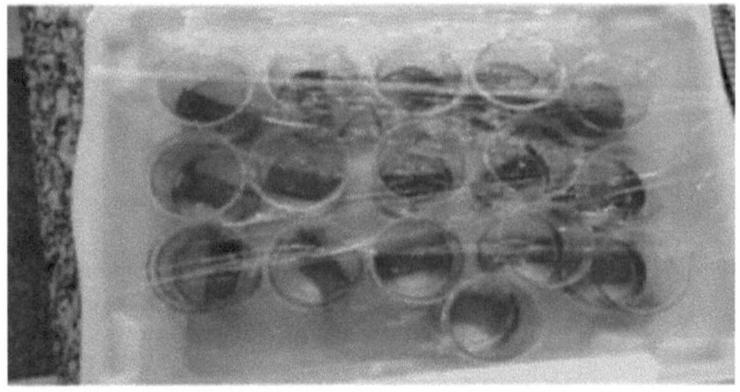

Figure 2 - Test specimens (1020 carbon steel plates / beakers 1 to 16/ Tab 1) immersed in saline solution (NaCl, 100ppm)

2.8 MASS LOSS METHOD

Weight loss measurements were taken every day during the four days of testing, at the same time of day, in order to study the loss of mass and the rate of corrosion in the parts. At the end of the four-day test, the specimens were removed from the solution, washed with distilled water and trichloroethylene and then weighed.

The percentage of inhibition efficiency was calculated from Equation 2.

(2)   Equation 2

$Ei = 100 \times \square\square - Pc\backslash Ps$

Where:

Ei is the percentage of inhibition efficiency;

Ps and Pc are the corrosion rates of the 1020 carbon steel plate in the absence and presence of the extract, respectively, in percent (GOMES *et al.*, 2005).

2.9 CORROSION RATE STUDY

The corrosion rate can be determined in several ways, the most common of which is the loss of mass per unit area per unit time. By measuring the mass of all the specimens before immersion in the sodium chloride solution and then removing them once a day at the same time to weigh them over the four days of testing, the corrosion rate can be assessed. The value of the total mass loss at the end of the time was divided by the time the tests lasted (96 hours), thus giving an idea of the average speed (rate) of corrosion.

3 RESULTS AND DISCUSSION

3.1 PROTEIN TEST

The results obtained for the protein test of Ora- Pro-Nobis with thorn (*Pereskia aculeata*) and without thorn (*Pereskia aculeata var. godseffiana*) are shown in Table 3, with the respective standard deviations. According to IAL (2008), a factor of 6.25 (other foods) was used to calculate the crude protein content.

Table 3- Protein content of Ora-Por-Nobis with thorn (*Pereskia aculeata)* and without thorn (*Pereskia aculeata var. godseffian)*

Sample	Sample weight (g)	Volume of HCl used in the titration (mL)	Crude Protein Content (percent m/m)
Mucilage ora- pro- nobis with Thorn	0,3359 (+/0,0358)	8,4 (+/ 1,0148)	43,72 (+/ 0,6288)
Thornless ora- pro- nobis mucilage	0,3940 (+/0,0222)	4,1 (+/ 0,3214)	19,94 (+/ 0,3523)

According to the data shown in Table 3, and confirming the theory of Santhiyathan, *et al.* (2005), ora-pro- nobis with a thorn should have the best anti-corrosion properties in all the samples.

anti-corrosive power on all MASS LOSS samples.

METHOD

The daily evaluations of mass loss during the four days analyzed are shown in Table 4.

Table 4 - Mass loss values of the 1020 carbon steel specimen **Mass (g).**

Sample	Applied material	Day 1	Day 2	Day 3	Day 4	Total loss (g)

A	Thornless ora-pro-nobis mucilage	65,6833 (+/ 3,3298)	65,6737 (+/ 3,3347)	65,6683 (+/ 3,3378)	65,6657 (+/ 3,3383)	0,0220 (+/ 0,0077)
B	Mucilage ora-pro-nobis with thorns	67,6253 (+/ 4,8114)	67,6243 (+/ 4,8112)	67,6197 (+/ 4,8062)	67,6163 (+/ 4,8103)	0,0110 (+/ 0,0013)
C	Thornless ora-pro-nobis mucilage with resin	66,1093 (+/ 0,9016)	66,0890 (+/ 0,9041)	66,0827 (+/ 0.9078)	66,0723 (+/ 0,9022)	0,1070 (+/ 0,0070)
D	Ora-pro-nobis mucilage with thorn and resin	65,9663 (+/ 2,4099)	65,9290 (+/ 2,4352)	65,9160 (+/ 2,4357)	65,9143 (+/ 2,4356)	0,0960 (+/ 0,0031)
E	Resin (control)	65,6270 (+/ 2,7388)	65,6183 (+/ 2,7487)	65,6117 (+/ 2,7446)	65,5423 (+/ 2,7035)	0,1410 (+/ 0,0481)
F	White	66,6980	66,6130	66,5970	66,5890	0,1660

According to the data obtained (Table 4), mass loss occurred for all the specimens, but in different quantities. As expected, the greatest loss of mass occurred in the specimen used as a blank, since it did not have any type of protection.

The lowest mass loss was seen in the test specimen where only the mucilage was applied (Sample B; 0.0110g). Based on the percentage inhibition efficiency of each mucilage evaluated (Table 5), it was found that the ora-pro-nobis mucilage with thorn (Sample B) obtained 93.57% efficiency through mass loss and Eq. 2.

Table 5 - Corrosion rates and corrosion inhibition efficiency of mucilages and resins applied to 1020 carbon steel plates

Samples	Corrosion rate (g/h)	Efficiency (%)
A	0,00023	86,75
B	0,00011	93,57
C	0,00112	35,34
D	0,00100	42,37
E	0,00147	14,86
F	0,00173	-

3.2 CORROSION RATE STUDY

Graph 1, constructed from the mass loss values of each sample in relation to time, makes it easier to visualize the corrosion rate in the 1020 carbon steel plate.

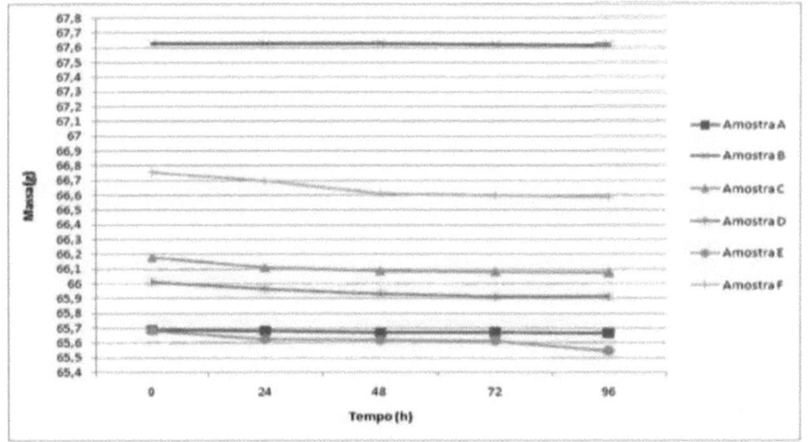

Graph 1 - Corrosion rate of 1020 carbon steel plate, coated with 4 different materials (ora-pro-nobis mucilage without thorn (Sample A), ora-pro-nobis mucilage with thorn (Sample B), ora-pro-nobis mucilage without thorn and with resin (Sample C), ora-pro-nobis mucilage with thorn and resin (Sample D), Resin (Sample E) and White (Sample F)) applied for four days Analyzing Graph 1, it can be seen that in the first 24 hours Samples C and E had a fairly

significant loss of mass, remaining almost constant over the next few hours. But at the end of the analysis, Sample E shows a more abrupt loss of mass, while C still has a low corrosion rate. Therefore, their application is not efficient during the time analyzed.

Samples D and F show a large loss of mass in the first 48 hours, after which the loss of mass continues, but less intensely. It can also be seen that Samples A and B do not show any abrupt drop during the 96 hours analyzed. The loss of mass in these samples is practically constant, which indicates a good result for the application of mucilage alone, when compared to the material applied using resin. Therefore, the 1020 carbon steel remained protected when only the mucilage *in natura* was applied.

According to Alencar, *et al.* (2013), an extract is efficient when the efficiency value reaches 70% or more in inhibiting corrosion, so the two mucilages applied are efficient. The highest efficiency was achieved by the mucilage from the ora-pro-nobis with thorn (Sample B; Table 5), reaching 93.57%. The second highest efficiency was the mucilage from the ora-pro-nobis without thorn (Sample A; Table 5), which reached 86.76%.

The mucilage of ora- pro-nobis with thorn showed a lower corrosion rate (Table 5), as verified by the mass loss tests (Table 4) and corrosion inhibition efficiency, this plant has a better potential for inhibiting corrosive processes due to its high protein content (Table 3).

4 CONCLUSION

For the *fresh* mucilages, the maximum inhibition efficiency was obtained with the ora-pro-nobis mucilage with thorn (Sample B, Table 5), reaching 93.57% efficiency, with a concentration of 300ppm of inhibitor in NaCl solution, at 100ppm. This can be attributed to the large quantities of proteins present in the extract, as these are extremely important substances in inhibiting corrosion.

With a satisfactory result (86.76% efficiency), the mucilage of the ora-pro-nobis without thorns (Sample A; Table 5), can also be considered a potential corrosion inhibitor.

According to the NP EN ISO 4628 standard, the resin based on ora-pro- nobis mucilage was not efficient, as it is classified as Ri5. Therefore, an in-depth study of the characteristics of mucilage should be carried out, as well as the components required for this resin to be efficient.

5.REFERENCES

AGEITEC - Embrapa Information and Technology Agency. **Plant characteristics**, 2010. Available at:http://www.agencia.cnptia.embrapa.br/gestor/mamona/rvore/CONT000h4rb0y9002wx7 ha0awymty4m5 2beo. html. Accessed on June 28, 2015.

AGOSTINI-COSTA, T. D. S. *et al.,* Carotenoids profile and total polyphenols in fruits of *Pereskia aculeata*

Miller. Revista Brasileira de Fruticultura, Jaboticabal, v. 34, n. 1, p. 234-238, 2012. ALENCAR, M. F. A. *et al,* Extracts from caatinga plants as corrosion inhibitors. 5th North-Northeast Congress of Chemistry, Fortaleza, 2013.

ALENCAR, M. F. A. *et al.,* **Extracts from caatinga plants as a corrosion inhibitor.** 5° Congresso norte-nordeste de quimica, Fortaleza, 2013. and phenolic acids: Role and biochemical activity in plants and human. Journal of Medicinal Plant Research, Lagos, v. 5, n. 31, p. 6697-6703, 2011. BOSSARDI, K.. Nanotechnology Applied to Surface Treatments for 1020 Carbon Steel as an Alternative to Zinc Phosphate. Federal University of Rio Grande do Sul, 2007.

CANGEMI, J. M.; SANTOS, A. M.; NETO, S. C., **The castor bean green revolution.** Quimica nova na escola, Sao Paulo, v. 32, n. 1, 2010.CPT. **Natural medicine - Babosa (Aloe spp.)**, 2015. Available at:

http://www.cpt.com.br/cursos- plantasmedicinais/artigos/medicina-natural- babosa- aloe-spp. Accessed on: June 30, 2015.

HEALING BY NATURE. **Ten reasons to use aloe vera**, 2013. Available at :http://www.curapelanatureza.com.br/2012/01/dez-   motivos-para-usar-babosa.html.

Accessed on: June 27, 2015.

FELIPE, M. B. M. C.; MACIEL, M. A. M.; MEDEIROS, S. R. B.; SILVA, D. R., **Aspectos gerais sobre corrosão e inibidores végétais.** Revista Virtual de Quimica, Rio Grande do Norte, v. 5, n. 4, 2013.FELIPE,

M. B. M. C.; MACIEL, M. A. M.;* MEDEIROS, S. R. B.; SILVA, D. R., **General aspects of corrosion and vegetable inhibitors.** Revista Virtual de Quimica, Rio Grande do Norte, v. 5, n. 4, 2013.

FRAUCHES-SANTOS, C. *et al*, **Corrosion and anticorrosive agents.** Revista Virtual de Quimica, Rio de Janeiro, v. 6, n. 2, 2013.

GENTIL, V. **Corrosion**. 5. ed. Rio de Janeiro: Livros Técnicos e Cientificos, 2007.

GOMES,F.F. *et al.,* Extract of Pectis oligocephala as an inhibitor of aluminum corrosion in 1M H2SO4. State University of Ceará, 2005.

GUNAVATHY, N. and MURUGAVEL, S. C. Corrosion IAL. Physico-chemical methods for food analysis. 4.ed. Sao Paulo: Adolfo Luttz Institute, 2008.

Inhibition Studies of Mild Steel in Acid Medium Using Musa Acuminata Fruit Peel Extract. E-Journal of Chemistry, 2012.

MARCOS, M., NUNES, U, B., **Against corrosion**, 2003.

Available at: http://www.inda.org.br/contra_corrosao.php. Accessed on November 15, 2015.

MARTINEZ, F. L. Plant tannins and their applications. University of Havana/Cuba.

Rio de Janeiro State University. October 1996.

MARTINS J.M., **Use of aloe vera in the repair of open wounds** caused surgically in dogs. **Federal University of Campina Grande, Center for Health and Rural Technology, Pernambuco, 2010.**

MAZZETTO, S. E., LOMONACO, D., MELE, G., Cashew nut **oil**: opportunities and challenges in the **context of industrial development and sustainability.** Revista Quimica Nova, Sao Paulo, v. 32, n. 3, 2009. OUR FUTURE

ROUBADO. **The incredible unconventional and super nutritious food plants in Brazil that we forget to eat,** 2011. Available at:

http://www.nossofuturoroubado.com.br/portal/as- amazing-unconventional-and-super-food-plants

nutritious-existing-in-brazil-that-we-forget-to-eat/. Accessed on October 21, 2015.

OLDHAM, K. B. and MYLAND, J. C.. Fundamentals of Electrochemical Science. California: Academic Press, Inc., 1994, 474p.

PALANISWAMY, N., **Corrosion inhibition of mild steel by ethanolic extracts of *Ricinus communis* leaves. Indian Journal of Chemical Technology, India, v. 12, 2005.**

PERES, R. S. Anticorrosive Properties of Tannin-Based Conversion Coats as a Pre-Treatment for 1020 Carbon Steel. Federal University of Rio Grande do Sul, 2010.

PERES, R. S., **Anticorrosive properties of tannin-based conversion layers as a pre-treatment for 1020 carbon steel.** Master's thesis. Federal University of Rio Grande do Sul, Porto Alegre, 2010.

RAHIM, A. A. *et al.,* Mangrove (Rhizophora apiculata) tannins: an eco-friendly rust converter. Corrosion Engineering, Science and Technology. 2011.

SATAPATHY, A. K. *et al.,* Corrosion inhibition by Justicia gendarussa plant extract in hydrochloric acid solution. Corros. Sci. 51 12 (2009) 2848.

SATHIYANATHAN, R. A. L. *et al,* Corrosion inhibition of mild steel by ethanolic extracts of Ricinus communis leaves. Indian Journal of Chemical Technology, India, v. 12, 2005.

SATHIYANATHAN, R. A. L.; MARUTHAMUTHU, S.; SELVANAYAGAM, M.; MOHAN AN, S.; SHARMA, S. K. and MUDHOO, A.. Green Corrosion Chemistry and Engineering; Opportunities and Challenges. Wiley-VCH Verlag GmbH&Co, 2012.

SOUZA, M. R. M. *el al.,* Potential of Ora-pronobis in the Diversification of Production Family Farming. Revista Brasileira De Agroecologia, Viçosa, v. 4, n. 2, 2009.

SOUZA, M. R. M.; CORREA, E. J. A.; GUIMARÂES, G.; PEREIRA, P. R. G., **O Potencial do Ora-pro-nobis na Diversificaçà da Produçao Agricola Familiar**. Revista Brasileira De Agroecologia, Viçosa, v. 4, n. 2, 2009.

STEVES, N., **The healing power of aloe**. 3.ed. Sao Paulo; Madras. 1999.

TECNICHE EUROPEE. TE 50180. Italy, 2011.